This planner belongs to:

_____

2026

January

S	M	T	W	T	F	S
				1	2	3
4	5	6	7	8	9	10
11	12	13	14	15	16	17
18	19	20	21	22	23	24
25	26	27	28	29	30	31

February

S	M	T	W	T	F	S
1	2	3	4	5	6	7
8	9	10	11	12	13	14
15	16	17	18	19	20	21
22	23	24	25	26	27	28

March

S	M	T	W	T	F	S
1	2	3	4	5	6	7
8	9	10	11	12	13	14
15	16	17	18	19	20	21
22	23	24	25	26	27	28
29	30	31				

April

S	M	T	W	T	F	S
			1	2	3	4
5	6	7	8	9	10	11
12	13	14	15	16	17	18
19	20	21	22	23	24	25
26	27	28	29	30		

May

S	M	T	W	T	F	S
					1	2
3	4	5	6	7	8	9
10	11	12	13	14	15	16
17	18	19	20	21	22	23
24	25	26	27	28	29	30
31						

June

S	M	T	W	T	F	S
	1	2	3	4	5	6
7	8	9	10	11	12	13
14	15	16	17	18	19	20
21	22	23	24	25	26	27
28	29	30				

July

S	M	T	W	T	F	S
			1	2	3	4
5	6	7	8	9	10	11
12	13	14	15	16	17	18
19	20	21	22	23	24	25
26	27	28	29	30	31	

August

S	M	T	W	T	F	S
						1
2	3	4	5	6	7	8
9	10	11	12	13	14	15
16	17	18	19	20	21	22
23	24	25	26	27	28	29
30	31					

September

S	M	T	W	T	F	S
		1	2	3	4	5
6	7	8	9	10	11	12
13	14	15	16	17	18	19
20	21	22	23	24	25	26
27	28	29	30			

October

S	M	T	W	T	F	S
				1	2	3
4	5	6	7	8	9	10
11	12	13	14	15	16	17
18	19	20	21	22	23	24
25	26	27	28	29	30	31

November

S	M	T	W	T	F	S
1	2	3	4	5	6	7
8	9	10	11	12	13	14
15	16	17	18	19	20	21
22	23	24	25	26	27	28
29	30					

December

S	M	T	W	T	F	S
		1	2	3	4	5
6	7	8	9	10	11	12
13	14	15	16	17	18	19
20	21	22	23	24	25	26
27	28	29	30	31		

Year in Pixels

	J	F	M	A	M	J	J	A	S	O	N	D
1.												
2.												
3.												
4.												
5.												
6.												
7.												
8.												
9.												
10.												
11.												
12.												
13.												
14.												
15.												
16.												
17.												
18.												
19.												
20.												
21.												
22.												
23.												
24.												
25.												
26.												
27.												
28.												
29.												
30.												
31.												

Color Codes

Notes

January

2026

MONDAY	TUESDAY	WEDNESDAY	THURSDAY
			1
5	6	7	8
12	13	14	15
19	20	21	22
26	27	28	29

January 2026

FRIDAY	SATURDAY	SUNDAY	NOTES
2	3	4	○
			○
			○
			○
			○
9	10	11	○
			○
			○
			○
16	17	18	○
			○
			○
			○
23	24	25	○
			○
			○
			○
			○
30	31		NOTES

February 2026

MONDAY	TUESDAY	WEDNESDAY	THURSDAY
2	3	4	5
9	10	11	12
16	17	18	19
23	24	25	26

February 2026

FRIDAY	SATURDAY	SUNDAY	NOTES
		1	○
			○
			○
			○
			○
6	7	8	○
			○
			○
			○
			○
13	14	15	○
			○
			○
			○
			○
20	21	22	○
			○
			○
			○
			○
27	28		NOTES

March

MONDAY	TUESDAY	WEDNESDAY	THURSDAY
2	3	4	5
9	10	11	12
16	17	18	19
23	24	25	26

March
2026

FRIDAY	SATURDAY	SUNDAY	NOTES
		1	○
			○
			○
			○
			○
6	7	8	○
			○
			○
			○
13	14	15	○
			○
			○
			○
			○
20	21	22	○
			○
			○
			○
			○
27	28	29	30
			31

April 2026

MONDAY	TUESDAY	WEDNESDAY	THURSDAY
		1	2
6	7	8	9
13	14	15	16
20	21	22	23
27	28	29	30

April

FRIDAY	SATURDAY	SUNDAY	NOTES
3	4	5	○
			○
			○
			○
			○
10	11	12	○
			○
			○
			○
			○
17	18	19	○
			○
			○
			○
			○
24	25	26	○
			○
			○
			○
			○
			NOTES

May

MONDAY	TUESDAY	WEDNESDAY	THURSDAY
4	5	6	7
11	12	13	14
18	19	20	21
25	26	27	28

May

FRIDAY	SATURDAY	SUNDAY	NOTES
1	2	3	○
			○
			○
			○
			○
8	9	10	○
			○
			○
			○
			○
15	16	17	○
			○
			○
			○
22	23	24	○
			○
			○
			○
			○
29	30	31	Notes

June

MONDAY	TUESDAY	WEDNESDAY	THURSDAY
1	2	3	4
8	9	10	11
15	16	17	18
22	23	24	25
29	30		

June

2026

FRIDAY	SATURDAY	SUNDAY	NOTES
5	6	7	○
			○
			○
			○
			○
12	13	14	○
			○
			○
			○
			○
19	20	21	○
			○
			○
			○
			○
26	27	28	○
			○
			○
			○
			○
			Notes

July

MONDAY	TUESDAY	WEDNESDAY	THURSDAY
		1	2
6	7	8	9
13	14	15	16
20	21	22	23
27	28	29	30

July 2026

FRIDAY	SATURDAY	SUNDAY	NOTES
3	4	5	○
			○
			○
			○
			○
10	11	12	○
			○
			○
			○
			○
17	18	19	○
			○
			○
			○
24	25	26	○
			○
			○
			○
			○
31			NOTES

August

MONDAY	TUESDAY	WEDNESDAY	THURSDAY
3	4	5	6
10	11	12	13
17	18	19	20
24	25	26	27

August

FRIDAY	SATURDAY	SUNDAY	NOTES
	1	2	○
			○
			○
			○
			○
7	8	9	○
			○
			○
			○
14	15	16	○
			○
			○
			○
			○
21	22	23	○
			○
			○
			○
			○
28	29	30	31

September

MONDAY	TUESDAY	WEDNESDAY	THURSDAY
	1	2	3
7	8	9	10
14	15	16	17
21	22	23	24
28	29	30	

September

FRIDAY	SATURDAY	SUNDAY	NOTES
4	5	6	○
			○
			○
			○
			○
11	12	13	○
			○
			○
			○
			○
18	19	20	○
			○
			○
			○
25	26	27	○
			○
			○
			○
			○
			NOTES

October 2026

MONDAY	TUESDAY	WEDNESDAY	THURSDAY
			1
5	6	7	8
12	13	14	15
19	20	21	22
26	27	28	29

October

2026

FRIDAY	SATURDAY	SUNDAY	NOTES
2	3	4	○
			○
			○
			○
			○
9	10	11	○
			○
			○
			○
			○
16	17	18	○
			○
			○
			○
23	24	25	○
			○
			○
			○
			○
30	31		NOTES

November

MONDAY	TUESDAY	WEDNESDAY	THURSDAY
2	3	4	5
9	10	11	12
16	17	18	19
23	24	25	26

November

2026

FRIDAY	SATURDAY	SUNDAY	NOTES
		1	○
			○
			○
			○
			○
6	7	8	○
			○
			○
			○
13	14	15	○
			○
			○
			○
			○
20	21	22	○
			○
			○
			○
			○
27	28	29	30

December 2026

MONDAY	TUESDAY	WEDNESDAY	THURSDAY
	1	2	3
7	8	9	10
14	15	16	17
21	22	23	24
28	29	30	31

December

FRIDAY	SATURDAY	SUNDAY	NOTES
4	5	6	○
			○
			○
			○
			○
11	12	13	○
			○
			○
			○
18	19	20	○
			○
			○
			○
			○
25	26	27	○
			○
			○
			○
			○
			NOTES

December
2025

01 MONDAY

02 TUESDAY

03 WEDNESDAY

04 FRIDAY

December
2025

05 FRIDAY

○
○
○
○
○
○
○
○

06 SATURDAY

○
○
○
○
○
○
○
○

07 SUNDAY

○
○
○
○
○
○
○
○

08 MONDAY

○
○
○
○
○

09 TUESDAY

10 WEDNESDAY

11 THURSDAY

12 FRIDAY

13 SATURDAY

14 SUNDAY

15 MONDAY

16 TUESDAY

17 WEDNESDAY

18 THURSDAY

19 FRIDAY

20 SATURDAY

21 SUNDAY

○
○
○
○
○
○
○
○

22 MONDAY

○
○
○
○
○
○
○
○

23 TUESDAY

○
○
○
○
○
○
○
○

24 WEDNESDAY

○
○
○
○
○

25 THURSDAY

26 FRIDAY

27 SATURDAY

28 SUNDAY

December
2025

29 MONDAY

○ _____
○ _____
○ _____
○ _____
○ _____
○ _____
○ _____
○ _____

30 TUESDAY

○ _____
○ _____
○ _____
○ _____
○ _____
○ _____
○ _____
○ _____

31 WEDNESDAY

○ _____
○ _____
○ _____
○ _____
○ _____
○ _____
○ _____
○ _____

NOTES

_____
_____
_____
_____

01 THURSDAY

○
○
○
○
○
○
○
○

02 FRIDAY

○
○
○
○
○
○
○
○

03 SATURDAY

○
○
○
○
○
○
○
○

04 SUNDAY

○
○
○
○
○

05 MONDAY

○
○
○
○
○
○
○
○

06 TUESDAY

○
○
○
○
○
○
○
○

07 WEDNESDAY

○
○
○
○
○
○
○
○

08 THURSDAY

○
○
○
○
○

09 FRIDAY

- ○
- ○
- ○
- ○
- ○
- ○
- ○
- ○

10 SATURDAY

- ○
- ○
- ○
- ○
- ○
- ○
- ○
- ○

11 SUNDAY

- ○
- ○
- ○
- ○
- ○
- ○
- ○
- ○

12 MONDAY

- ○
- ○
- ○
- ○
- ○

13 TUESDAY

-
-
-
-
-
-
-
-

14 WEDNESDAY

-
-
-
-
-
-
-
-

15 THURSDAY

-
-
-
-
-
-
-
-

16 FRIDAY

-
-
-
-
-

17 SATURDAY

18 SUNDAY

19 MONDAY

20 TUESDAY

21 WEDNESDAY

22 THURSDAY

23 FRIDAY

24 SATURDAY

25 SUNDAY

26 MONDAY

27 TUESDAY

28 WEDNESDAY

29 THURSDAY

- ○
- ○
- ○
- ○
- ○
- ○
- ○
- ○

30 FRIDAY

- ○
- ○
- ○
- ○
- ○
- ○
- ○
- ○

31 SATURDAY

- ○
- ○
- ○
- ○
- ○
- ○
- ○
- ○

NOTES

February
2026

01 SUNDAY
○
○
○
○
○
○
○
○

02 MONDAY
○
○
○
○
○
○
○
○

03 TUESDAY
○
○
○
○
○
○
○
○

04 WEDNESDAY
○
○
○
○
○

February
2026

05 THURSDAY

- ○
- ○
- ○
- ○
- ○
- ○
- ○
- ○

06 FRIDAY

- ○
- ○
- ○
- ○
- ○
- ○
- ○
- ○

07 SATURDAY

- ○
- ○
- ○
- ○
- ○
- ○
- ○
- ○

08 SUNDAY

- ○
- ○
- ○
- ○
- ○

February
2026

09 MONDAY

○
○
○
○
○
○
○
○

10 TUESDAY

○
○
○
○
○
○
○
○

11 WEDNESDAY

○
○
○
○
○
○
○
○

12 THURSDAY

○
○
○
○
○

13 FRIDAY

○ —————————————————————————————
○ —————————————————————————————
○ —————————————————————————————
○ —————————————————————————————
○ —————————————————————————————
○ —————————————————————————————
○ —————————————————————————————
○ —————————————————————————————

14 SATURDAY

○ —————————————————————————————
○ —————————————————————————————
○ —————————————————————————————
○ —————————————————————————————
○ —————————————————————————————
○ —————————————————————————————
○ —————————————————————————————
○ —————————————————————————————

15 SUNDAY

○ —————————————————————————————
○ —————————————————————————————
○ —————————————————————————————
○ —————————————————————————————
○ —————————————————————————————
○ —————————————————————————————
○ —————————————————————————————
○ —————————————————————————————

16 MONDAY

○ —————————————————————————————
○ —————————————————————————————
○ —————————————————————————————
○ —————————————————————————————
○ —————————————————————————————

17 TUESDAY

18 WEDNESDAY

19 THURSDAY

20 FRIDAY

February
2026

21 SATURDAY

22 SUNDAY

23 MONDAY

24 TUESDAY

February 2026

25 WEDNESDAY

○
○
○
○
○
○
○
○

26 THURSDAY

○
○
○
○
○
○
○
○

27 FRIDAY

○
○
○
○
○
○
○
○

28 SATURDAY

○
○
○
○
○

01 SUNDAY

○ _____
○ _____
○ _____
○ _____
○ _____
○ _____
○ _____
○ _____

02 MONDAY

○ _____
○ _____
○ _____
○ _____
○ _____
○ _____
○ _____
○ _____

03 TUESDAY

○ _____
○ _____
○ _____
○ _____
○ _____
○ _____
○ _____
○ _____

04 WEDNESDAY

○ _____
○ _____
○ _____
○ _____
○ _____

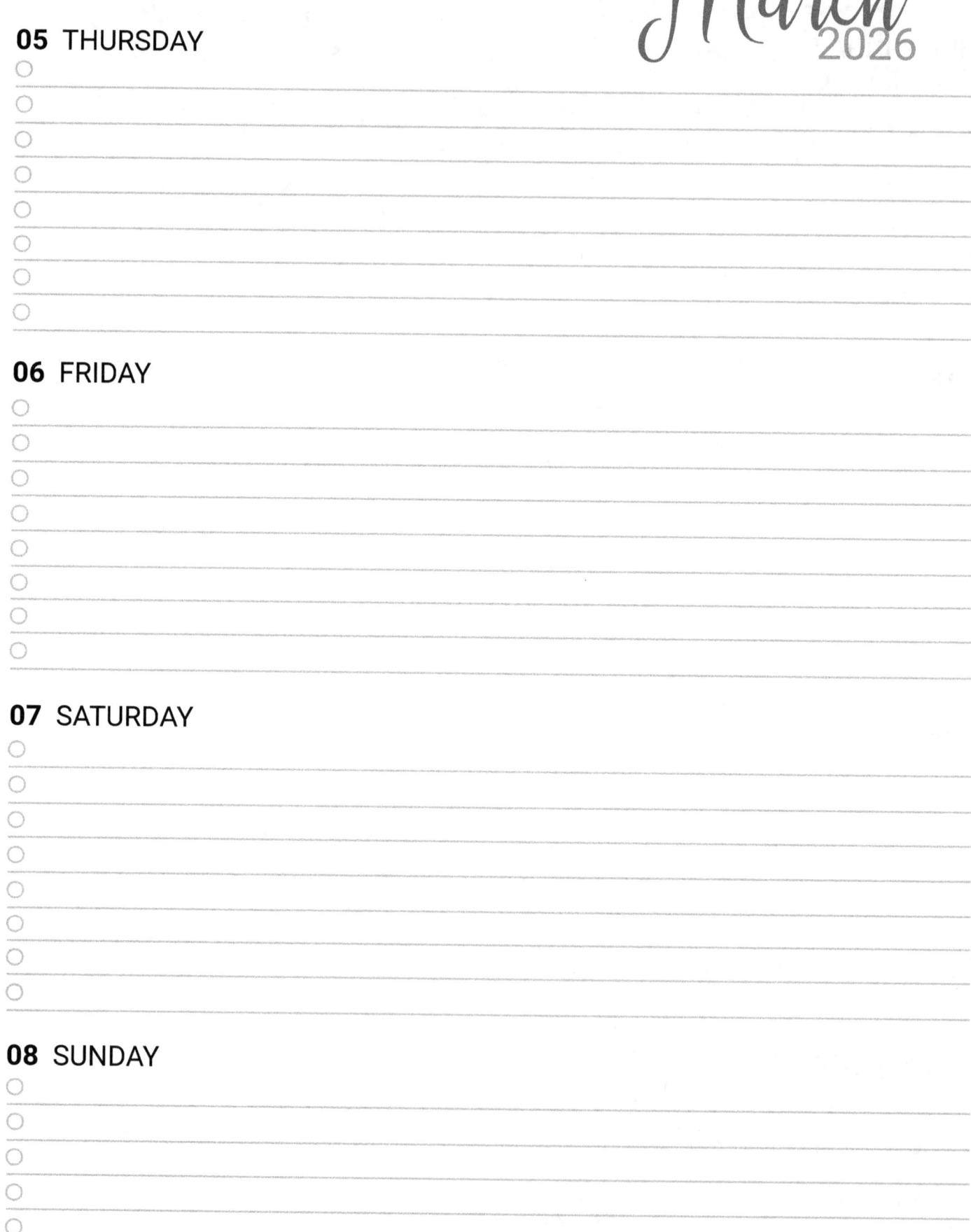

March
2026

05 THURSDAY

○
○
○
○
○
○
○
○

06 FRIDAY

○
○
○
○
○
○
○
○

07 SATURDAY

○
○
○
○
○
○
○
○

08 SUNDAY

○
○
○
○
○

09 MONDAY

10 TUESDAY

11 WEDNESDAY

12 THURSDAY

13 FRIDAY

○
○
○
○
○
○
○
○

14 SATURDAY

○
○
○
○
○
○
○
○

15 SUNDAY

○
○
○
○
○
○
○
○

16 MONDAY

○
○
○
○
○

17 TUESDAY

18 WEDNESDAY

19 THURSDAY

20 FRIDAY

21 SATURDAY

22 SUNDAY

23 MONDAY

24 TUESDAY

March 2026

25 WEDNESDAY

26 THURSDAY

27 FRIDAY

28 SATURDAY

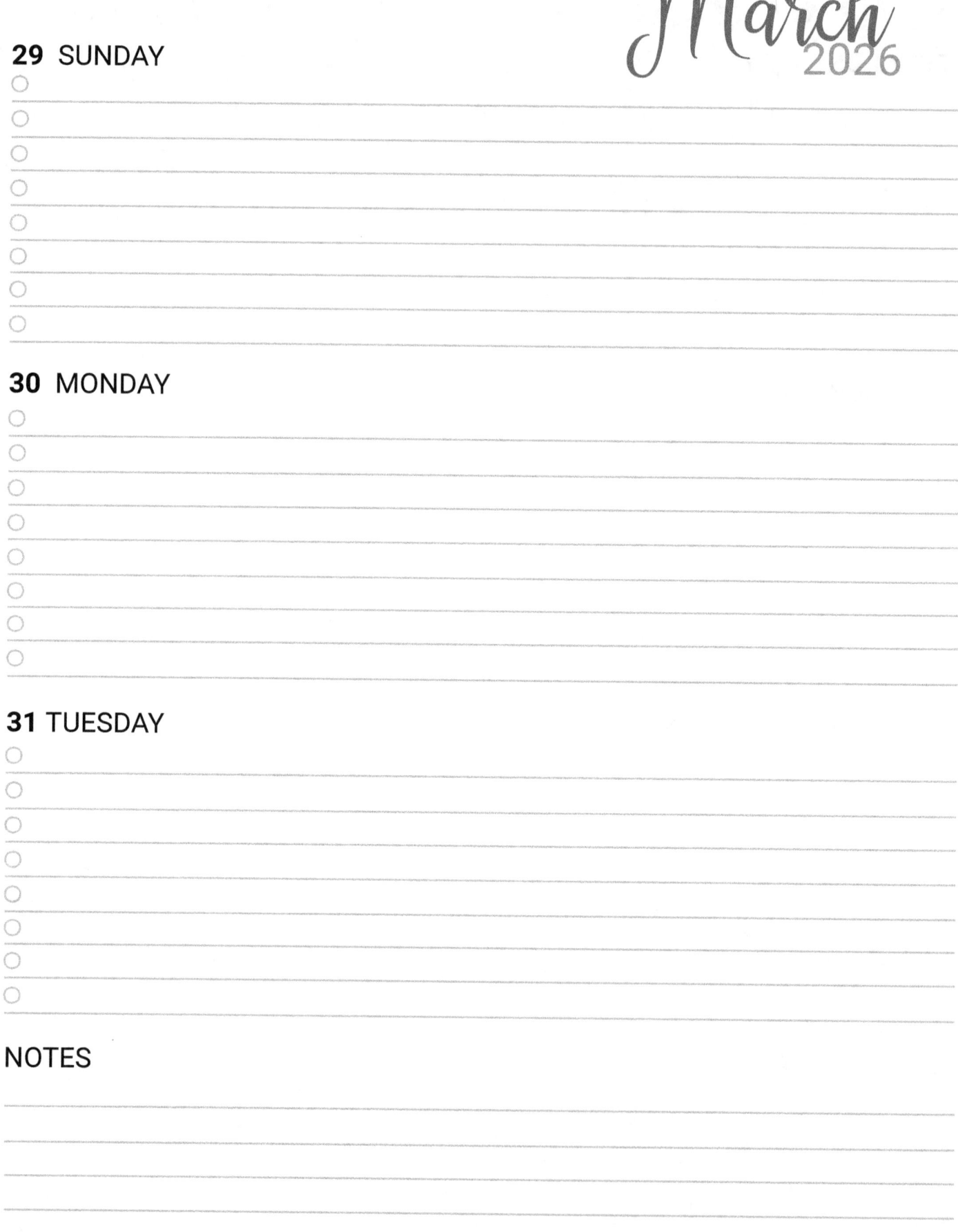

March 2026

29 SUNDAY

30 MONDAY

31 TUESDAY

NOTES

01 WEDNESDAY

○ _____
○ _____
○ _____
○ _____
○ _____
○ _____
○ _____
○ _____

02 THURSDAY

○ _____
○ _____
○ _____
○ _____
○ _____
○ _____
○ _____
○ _____

03 FRIDAY

○ _____
○ _____
○ _____
○ _____
○ _____
○ _____
○ _____
○ _____

04 SATURDAY

○ _____
○ _____
○ _____
○ _____
○ _____

April
2026

05 SUNDAY

06 MONDAY

07 TUESDAY

08 WEDNESDAY

April 2026

09 THURSDAY

10 FRIDAY

11 SATURDAY

12 SUNDAY

13 MONDAY

○
○
○
○
○
○
○
○

14 TUESDAY

○
○
○
○
○
○
○
○

15 WEDNESDAY

○
○
○
○
○
○
○
○

16 THURSDAY

○
○
○
○
○

April
2026

17 FRIDAY

○
○
○
○
○
○
○
○

18 SATURDAY

○
○
○
○
○
○
○
○

19 SUNDAY

○
○
○
○
○
○
○
○

20 MONDAY

○
○
○
○
○

21 TUESDAY

22 WEDNESDAY

23 THURSDAY

24 FRIDAY

25 SATURDAY

○
○
○
○
○
○
○
○

26 SUNDAY

○
○
○
○
○
○
○
○

27 MONDAY

○
○
○
○
○
○
○
○

28 TUESDAY

○
○
○
○
○

29 WEDNESDAY

○ _____
○ _____
○ _____
○ _____
○ _____
○ _____
○ _____
○ _____

30 THURSDAY

○ _____
○ _____
○ _____
○ _____
○ _____
○ _____
○ _____
○ _____

NOTES

_____
_____
_____
_____
_____
_____
_____
_____
_____
_____
_____
_____

May
2026

01 FRIDAY

○
○
○
○
○
○
○
○

02 SATURDAY

○
○
○
○
○
○
○
○

03 SUNDAY

○
○
○
○
○
○
○
○

04 MONDAY

○
○
○
○
○

May
2026

05 TUESDAY

○
○
○
○
○
○
○
○

06 WEDNESDAY

○
○
○
○
○
○
○
○

07 THURSDAY

○
○
○
○
○
○
○
○

08 FRIDAY

○
○
○
○
○

09 SATURDAY

10 SUNDAY

11 MONDAY

12 TUESDAY

13 WEDNESDAY

○
○
○
○
○
○
○
○

14 THURSDAY

○
○
○
○
○
○
○
○

15 FRIDAY

○
○
○
○
○
○
○
○

16 SATURDAY

○
○
○
○
○

May
2026

17 SUNDAY

○
○
○
○
○
○
○
○

18 MONDAY

○
○
○
○
○
○
○
○

19 TUESDAY

○
○
○
○
○
○
○
○

20 WEDNESDAY

○
○
○
○
○

21 THURSDAY

22 FRIDAY

23 SATURDAY

24 SUNDAY

May
2026

25 MONDAY

○
○
○
○
○
○
○
○

26 TUESDAY

○
○
○
○
○
○
○
○

27 WEDNESDAY

○
○
○
○
○
○
○
○

28 THURSDAY

○
○
○
○
○

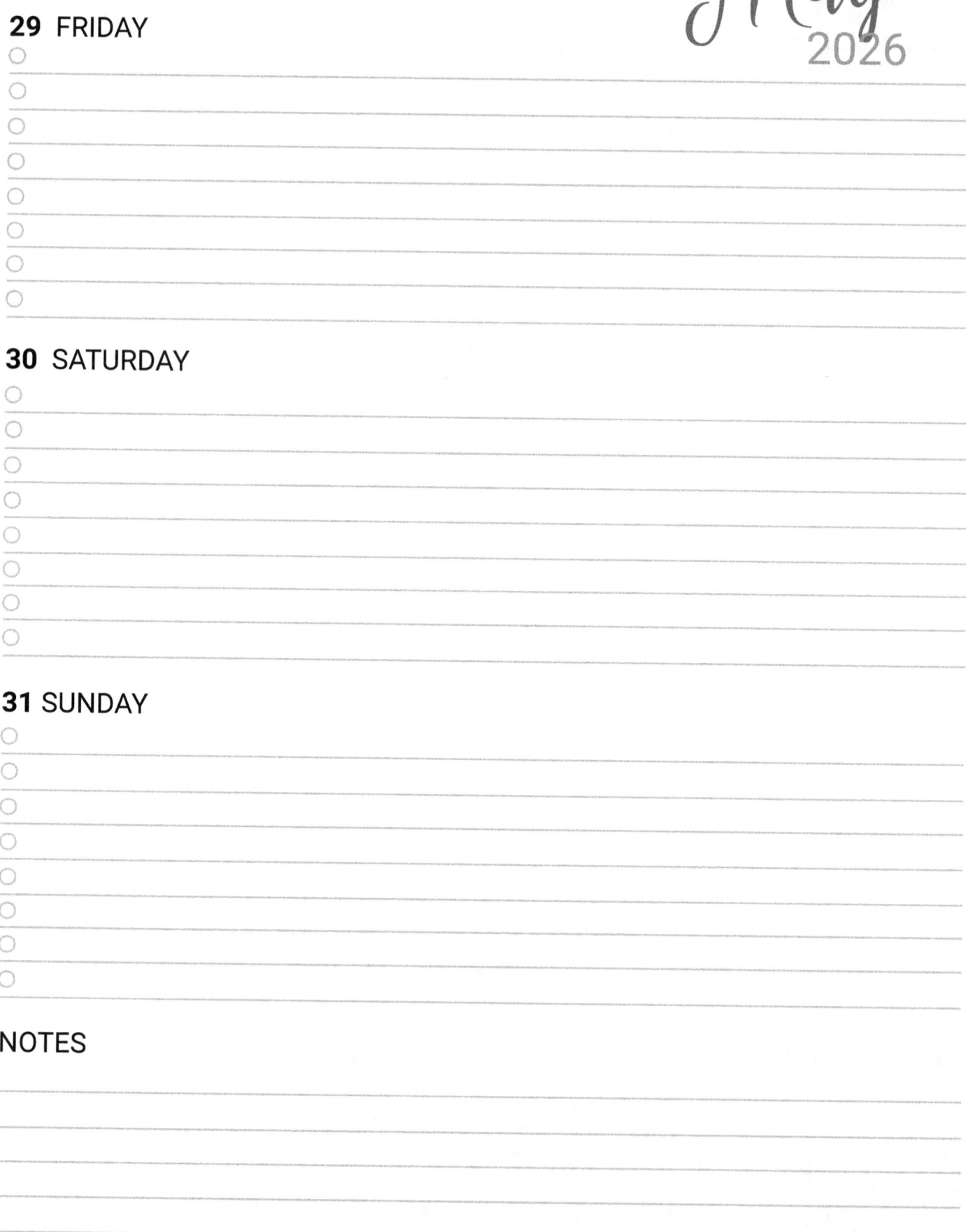

May
2026

29 FRIDAY

30 SATURDAY

31 SUNDAY

NOTES

June
2026

01 MONDAY
- ○
- ○
- ○
- ○
- ○
- ○
- ○
- ○

02 TUESDAY
- ○
- ○
- ○
- ○
- ○
- ○
- ○
- ○

03 WEDNESDAY
- ○
- ○
- ○
- ○
- ○
- ○
- ○
- ○

04 THURSDAY
- ○
- ○
- ○
- ○
- ○

June 2026

05 FRIDAY

○
○
○
○
○
○
○
○

06 SATURDAY

○
○
○
○
○
○
○
○

07 SUNDAY

○
○
○
○
○
○
○
○

08 MONDAY

○
○
○
○
○

09 TUESDAY

10 WEDNESDAY

11 THURSDAY

12 FRIDAY

13 SATURDAY

○
○
○
○
○
○
○
○

14 SUNDAY

○
○
○
○
○
○
○
○

15 MONDAY

○
○
○
○
○
○
○
○

16 TUESDAY

○
○
○
○
○

17 WEDNESDAY

18 THURSDAY

19 FRIDAY

20 SATURDAY

June 2026

21 SUNDAY

22 MONDAY

23 TUESDAY

24 WEDNESDAY

25 THURSDAY

○
○
○
○
○
○
○
○

26 FRIDAY

○
○
○
○
○
○
○
○

27 SATURDAY

○
○
○
○
○
○
○
○

28 SUNDAY

○
○
○
○
○

June
2026

29 MONDAY

○
○
○
○
○
○
○
○

30 TUESDAY

○
○
○
○
○
○
○
○

NOTES

July 2026

01 WEDNESDAY

02 THURSDAY

03 FRIDAY

04 SATURDAY

July
2026

05 SUNDAY

06 MONDAY

07 TUESDAY

08 WEDNESDAY

July
2026

09 THURSDAY

○ _____
○ _____
○ _____
○ _____
○ _____
○ _____
○ _____
○ _____

10 FRIDAY

○ _____
○ _____
○ _____
○ _____
○ _____
○ _____
○ _____
○ _____

11 SATURDAY

○ _____
○ _____
○ _____
○ _____
○ _____
○ _____
○ _____
○ _____

12 SUNDAY

○ _____
○ _____
○ _____
○ _____
○ _____

13 MONDAY

14 TUESDAY

15 WEDNESDAY

16 THURSDAY

17 FRIDAY

○
○
○
○
○
○
○
○

18 SATURDAY

○
○
○
○
○
○
○
○

19 SUNDAY

○
○
○
○
○
○
○
○

20 MONDAY

○
○
○
○
○

21 TUESDAY

○

○

○

○

○

○

○

○

22 WEDNESDAY

○

○

○

○

○

○

○

○

23 THURSDAY

○

○

○

○

○

○

○

○

24 FRIDAY

○

○

○

○

○

25 SATURDAY

26 SUNDAY

27 MONDAY

28 TUESDAY

29 WEDNESDAY

○
○
○
○
○
○
○
○

30 THURSDAY

○
○
○
○
○
○
○
○

31 FRIDAY

○
○
○
○
○
○
○

NOTES

August
2026

01 SATURDAY
○
○
○
○
○
○
○
○

02 SUNDAY
○
○
○
○
○
○
○
○

03 MONDAY
○
○
○
○
○
○
○
○

04 TUESDAY
○
○
○
○
○

August
2026

05 WEDNESDAY

06 THURSDAY

07 FRIDAY

08 SATURDAY

09 SUNDAY

○ _____
○ _____
○ _____
○ _____
○ _____
○ _____
○ _____
○ _____

10 MONDAY

○ _____
○ _____
○ _____
○ _____
○ _____
○ _____
○ _____
○ _____

11 TUESDAY

○ _____
○ _____
○ _____
○ _____
○ _____
○ _____
○ _____

12 WEDNESDAY

○ _____
○ _____
○ _____
○ _____
○ _____

13 THURSDAY

14 FRIDAY

15 SATURDAY

16 SUNDAY

17 MONDAY

○
○
○
○
○
○
○
○

18 TUESDAY

○
○
○
○
○
○
○
○

19 WEDNESDAY

○
○
○
○
○
○
○
○

20 THURSDAY

○
○
○
○
○

21 FRIDAY

○
○
○
○
○
○
○
○

22 SATURDAY

○
○
○
○
○
○
○
○

23 SUNDAY

○
○
○
○
○
○
○
○

24 MONDAY

○
○
○
○
○

25 TUESDAY

○
○
○
○
○
○
○
○

26 WEDNESDAY

○
○
○
○
○
○
○
○

27 THURSDAY

○
○
○
○
○
○
○
○

28 FRIDAY

○
○
○
○
○

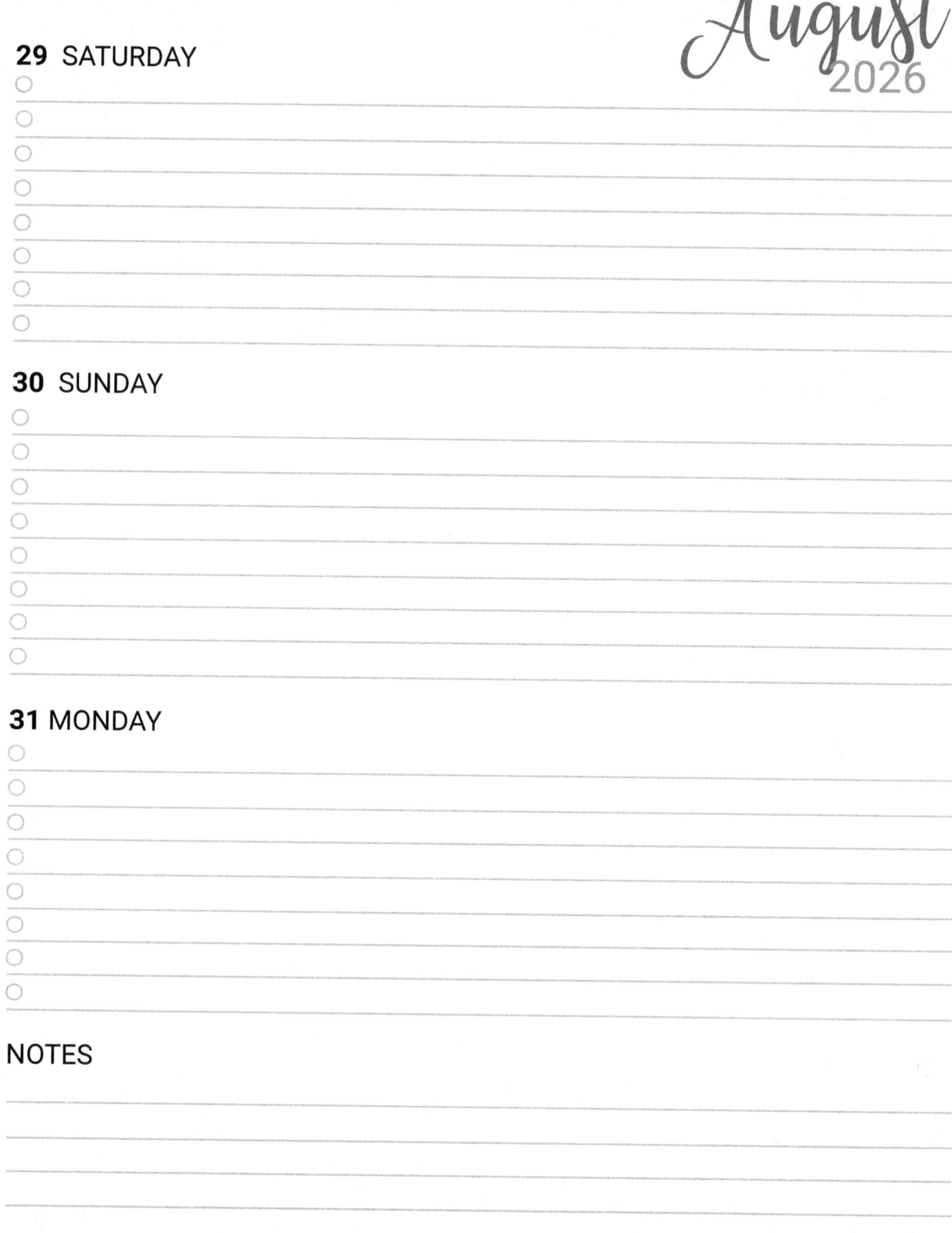

29 SATURDAY

30 SUNDAY

31 MONDAY

NOTES

01 TUESDAY

○ _____
○ _____
○ _____
○ _____
○ _____
○ _____
○ _____
○ _____

02 WEDNESDAY

○ _____
○ _____
○ _____
○ _____
○ _____
○ _____
○ _____
○ _____

03 THURSDAY

○ _____
○ _____
○ _____
○ _____
○ _____
○ _____
○ _____
○ _____

04 FRIDAY

○ _____
○ _____
○ _____
○ _____
○ _____

05 SATURDAY

06 SUNDAY

07 MONDAY

08 TUESDAY

09 WEDNESDAY

- ○
- ○
- ○
- ○
- ○
- ○
- ○
- ○

10 THURSDAY

- ○
- ○
- ○
- ○
- ○
- ○
- ○
- ○

11 FRIDAY

- ○
- ○
- ○
- ○
- ○
- ○
- ○
- ○

12 SATURDAY

- ○
- ○
- ○
- ○
- ○

13 SUNDAY

○
○
○
○
○
○
○
○

14 MONDAY

○
○
○
○
○
○
○
○

15 TUESDAY

○
○
○
○
○
○
○
○

16 WEDNESDAY

○
○
○
○
○

17 THURSDAY

○
○
○
○
○
○
○
○

18 FRIDAY

○
○
○
○
○
○
○
○

19 SATURDAY

○
○
○
○
○
○
○
○

20 SUNDAY

○
○
○
○
○

21 MONDAY
○
○
○
○
○
○
○
○

22 TUESDAY
○
○
○
○
○
○
○
○

23 WEDNESDAY
○
○
○
○
○
○
○
○

24 THURSDAY
○
○
○
○
○

25 FRIDAY

26 SATURDAY

27 SUNDAY

28 MONDAY

29 TUESDAY

○ _____
○ _____
○ _____
○ _____
○ _____
○ _____
○ _____
○ _____

30 WEDNESDAY

○ _____
○ _____
○ _____
○ _____
○ _____
○ _____
○ _____
○ _____

NOTES

_____
_____
_____
_____
_____
_____
_____
_____
_____
_____
_____
_____

October
2026

01 THURSDAY

02 FRIDAY

03 SATURDAY

04 SUNDAY

October
2026

05 MONDAY

06 TUESDAY

07 WEDNESDAY

08 THURSDAY

October
2026

09 FRIDAY

10 SATURDAY

11 SUNDAY

12 MONDAY

13 TUESDAY

14 WEDNESDAY

15 THURSDAY

16 FRIDAY

17 SATURDAY

18 SUNDAY

19 MONDAY

20 TUESDAY

21 WEDNESDAY

22 THURSDAY

23 FRIDAY

24 SATURDAY

25 SUNDAY

26 MONDAY

27 TUESDAY

28 WEDNESDAY

29 THURSDAY

○ _____
○ _____
○ _____
○ _____
○ _____
○ _____
○ _____
○ _____

30 FRIDAY

○ _____
○ _____
○ _____
○ _____
○ _____
○ _____
○ _____
○ _____

31 SATURDAY

○ _____
○ _____
○ _____
○ _____
○ _____
○ _____
○ _____
○ _____

NOTES

_____
_____
_____
_____

November
2026

01 SUNDAY

02 MONDAY

03 TUESDAY

04 WEDNESDAY

November
2026

05 THURSDAY

06 FRIDAY

07 SATURDAY

08 SUNDAY

09 MONDAY

○ _____
○ _____
○ _____
○ _____
○ _____
○ _____
○ _____
○ _____

10 TUESDAY

○ _____
○ _____
○ _____
○ _____
○ _____
○ _____
○ _____
○ _____

11 WEDNESDAY

○ _____
○ _____
○ _____
○ _____
○ _____
○ _____
○ _____
○ _____

12 THURSDAY

○ _____
○ _____
○ _____
○ _____
○ _____

13 FRIDAY

○
○
○
○
○
○
○

14 SATURDAY

○
○
○
○
○
○
○
○

15 SUNDAY

○
○
○
○
○
○
○
○

16 MONDAY

○
○
○
○
○

17 TUESDAY

18 WEDNESDAY

19 THURSDAY

20 FRIDAY

21 SATURDAY

22 SUNDAY

23 MONDAY

24 TUESDAY

25 WEDNESDAY

○ _____
○ _____
○ _____
○ _____
○ _____
○ _____
○ _____
○ _____

26 THURSDAY

○ _____
○ _____
○ _____
○ _____
○ _____
○ _____
○ _____
○ _____

27 FRIDAY

○ _____
○ _____
○ _____
○ _____
○ _____
○ _____
○ _____
○ _____

28 SATURDAY

○ _____
○ _____
○ _____
○ _____
○ _____

29 SUNDAY

○
○
○
○
○
○
○
○

30 MONDAY

○
○
○
○
○
○
○
○

NOTES

December
2026

01 TUESDAY
○
○
○
○
○
○
○
○

02 WEDNESDAY
○
○
○
○
○
○
○
○

03 THURSDAY
○
○
○
○
○
○
○
○

04 FRIDAY
○
○
○
○
○

December
2026

05 SATURDAY

06 SUNDAY

07 MONDAY

08 TUESDAY

December
2026

09 WEDNESDAY

○
○
○
○
○
○
○
○

10 THURSDAY

○
○
○
○
○
○
○
○

11 FRIDAY

○
○
○
○
○
○
○
○

12 SATURDAY

○
○
○
○
○

13 SUNDAY

14 MONDAY

15 TUESDAY

16 WEDNESDAY

17 THURSDAY

18 FRIDAY

19 SATURDAY

20 SUNDAY

21 MONDAY

22 TUESDAY

23 WEDNESDAY

24 THURSDAY

25 FRIDAY

○
○
○
○
○
○
○
○

26 SATURDAY

○
○
○
○
○
○
○
○

27 SUNDAY

○
○
○
○
○
○
○
○

28 MONDAY

○
○
○
○
○

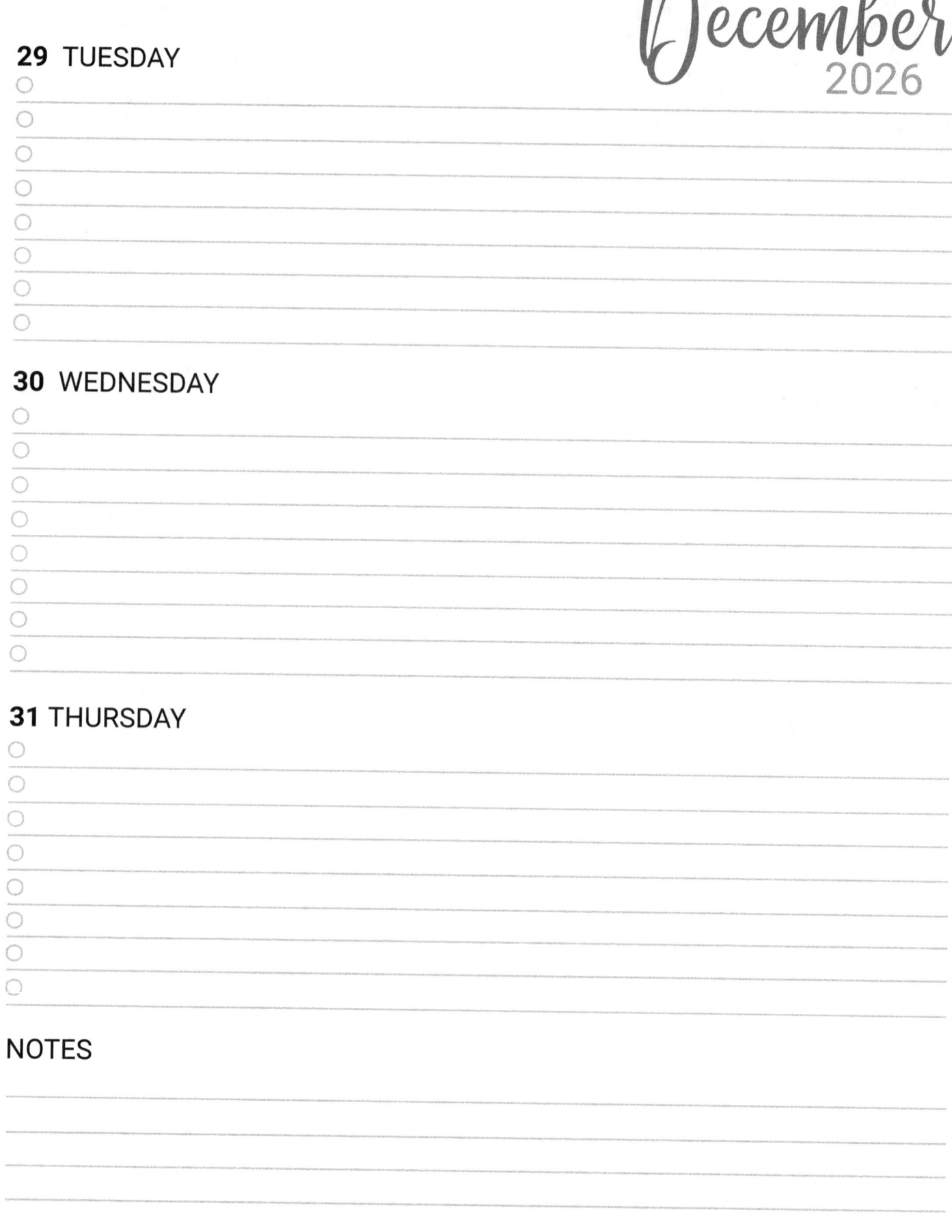

December
2026

29 TUESDAY

30 WEDNESDAY

31 THURSDAY

NOTES